BACK IN MOTION
Understanding the Healthiest You

Brian Narelle & Robert W. Adams, DC, DABCN
Board Certified in Chiropractic Neurology
Forward by Prof. Ted Carrick

Back In Motion
Understanding The Healthiest You

ISBN: 978-0-9991747-2-2

Limits of Liability/Disclaimer of Warranty

The following information is either from the medical literature or based on my years of clinical experience. My recommendations are for your information only and do not represent medical advice. Please discuss any proposed changes to your medical and chiropractic care with your trusted health care providers. Never disregard the advice of your trusted providers or delay seeking treatment because of the information that you have read or images you have seen in this book.
Any views in this book or opinions presented are solely those of Dr. Robert Adams and do not represent those of the publisher.

1st Printing January 2018
Printed in the United States of America

To better health from:

Dr. Robert W. Adams shares current evidence-based healthcare information about low back injury in an entertaining easy to read format. It is easy to read but it is not light on information. The cartoon effect complements the understanding of a difficult reality of human fragility. Dr. Adams leads us on a journey that most of us will experience sometime during our life. It is a terrible reality that back pain is a curse and this book promotes an understanding of why it happens to us and what we might do to help ourselves when it does. The book is complete with easy-to-follow exercises that are logical, realistic and simple to follow. All in all, the topic represents an epidemic of disability that has a severe physical and financial consequence on our society. Everyone will have the time to read this book, and prepare themselves for the inevitability of back pain and be familiar with what to do when it happens. I recommend it without equivocation and am very pleased that the complexity of pain in the nervous system can be illustrated so comprehensively. This is a worthwhile page turner.

Prof. Frederick R. Carrick, DC, PhD, MS-HPed
Roland Blaauw Professor of Neurology and
Senior Research Fellow BCMHR in association with
University of Cambridge

DEDICATION

This book, Back in Motion, is dedicated to my wonderful and patient wife Jamie. She has been in my life through my undergraduate university experience, Chiropractic education, post doctoral studies, and board certification in Clinical Neurology. She has kindly supported my pursuits to grasp the depth, breadth and clinical applications of Neurology. Her quiet strength has been an invaluable resource and motivation.

Robert W. Adams

BACK IN MOTION
Understanding the Healthiest You

One day you bend over to do something you've done a million times before...

and YOW! The pain!!

After all...you thought you were strong.

If a joint is locked up, contributing to weakened muscle and decreased feedback from movement receptors, all the exercise in the world is only going to make the unrestricted parts stronger as you await impending disaster.

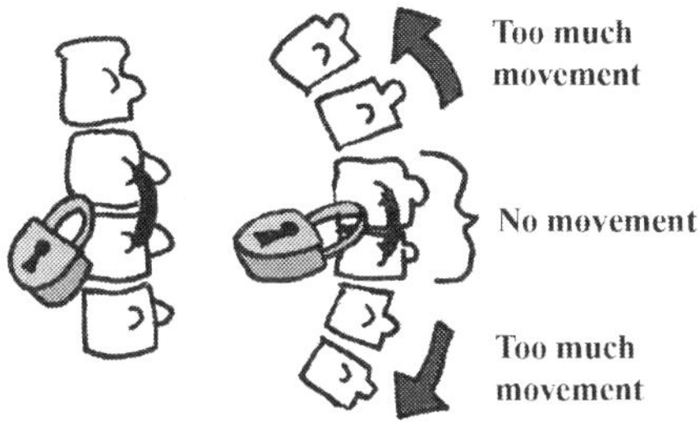

Movement receptors in the vertebral joints...

as well as in the surrounding muscle and connective tissues need stimulation.

Exercise won't unlock it. Sometimes a specific force is required to restore range of motion and movement receptor activation (mechanoreceptors)...

To reestablish communication with the brain, which in turn stimulates protein production in the muscle.

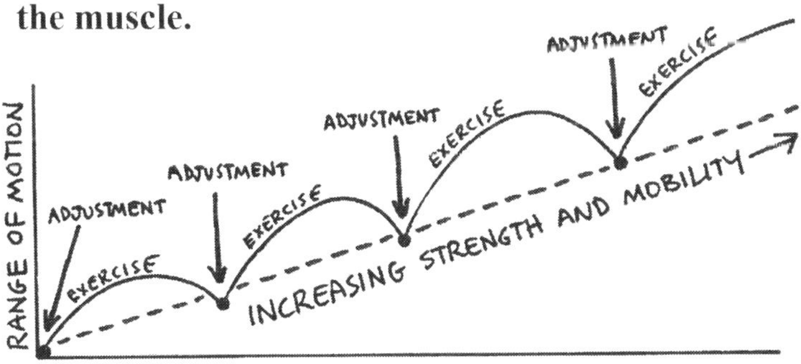

This is where chiropractic intervention becomes critical.

Weak muscles and frozen joints are impediments to restoring the stabilizing capacity of the nervous system and a spine with proper alignment and curvature

Chiropractic intervention is a about much more than just reestablishing alignment. Increasing range of motion is critical to establishing open communication to and from the brain, facilitating both volitional and compensatory muscle stimulation.

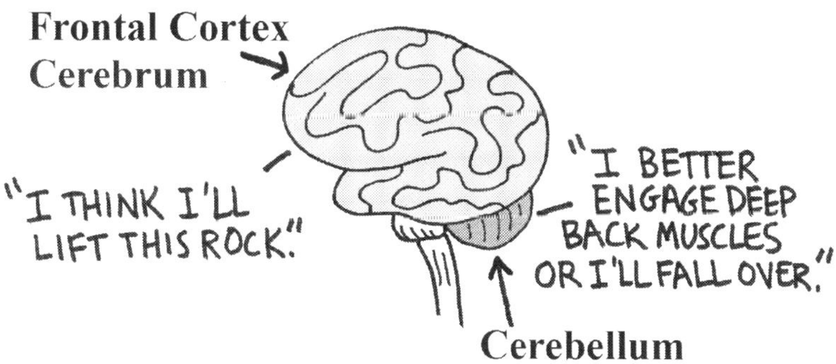

All at the same time... 

A brief sidetrack as physics meets physiology...

Plutonium 239 has a half life of 24,100 years.

Muscle protein has a half life of only 6 to ten days!

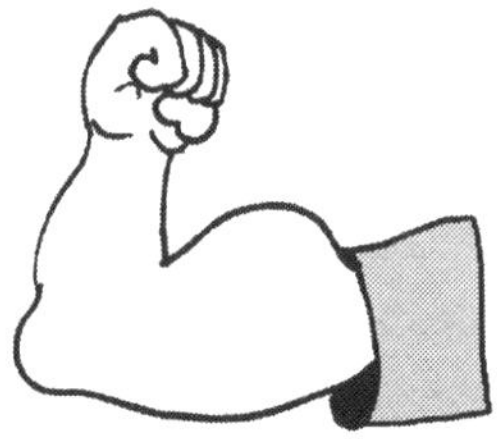

It's a...

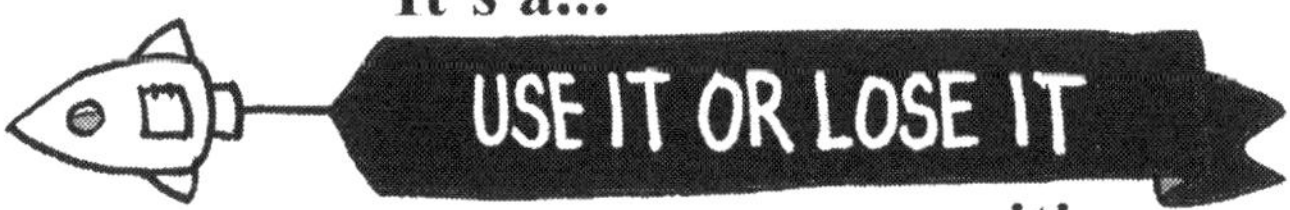

...proposition.

Your brain only knows about muscles if they are activated.

Immobility = No stimulation = No protein production
In 6-10 days of immobility you can lose 1/2 of unused muscle.

And it's all downhill from there.

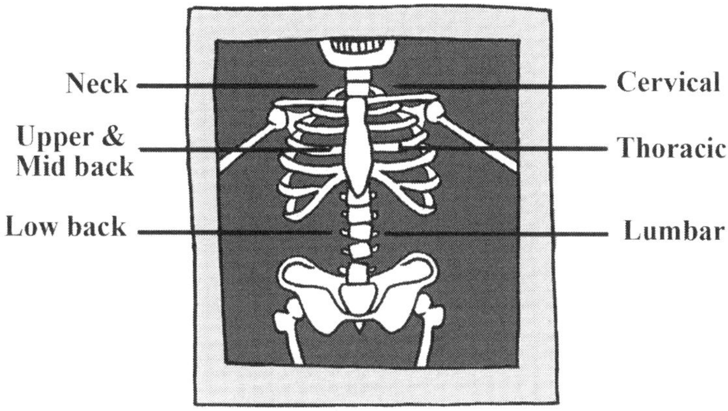

Labels can provide useful distinctions that aid in communication but it's important to remember that your body is one unified organism. Everything is connected.

And now a word from your CERELLUM *

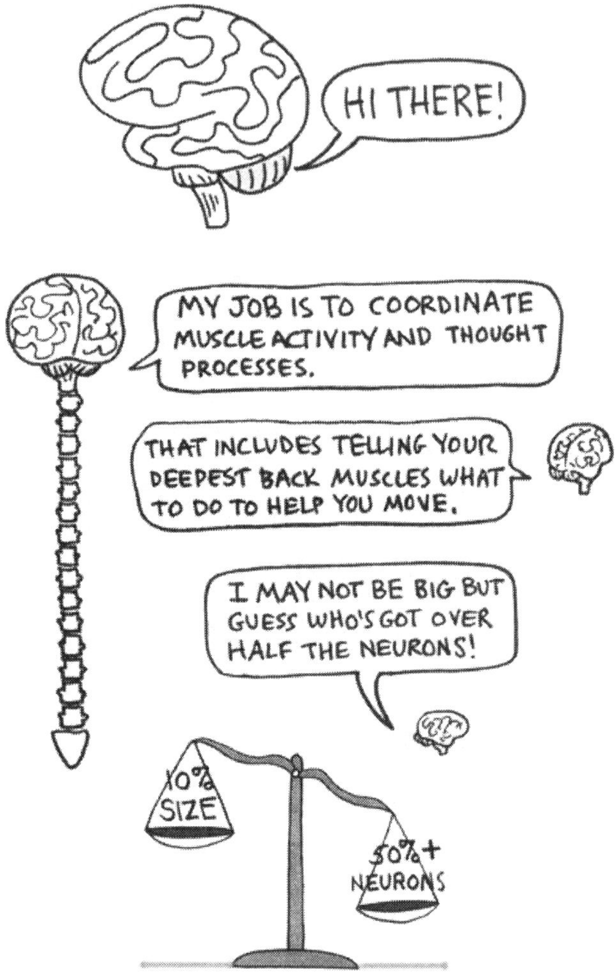

*A large portion of the brain, serving to coordinate voluntary movements, posture and balance in humans, being in back of and below the cerebrum and consisting of two lateral lobes and a central lobe.

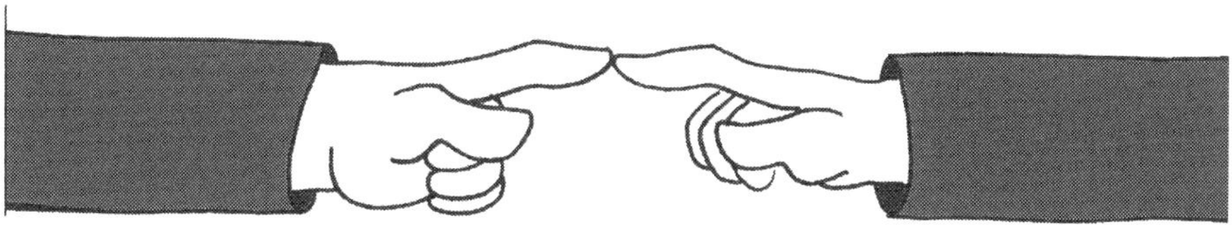

EVERYTHING'S CONNECTED

Grab rock and lift it up

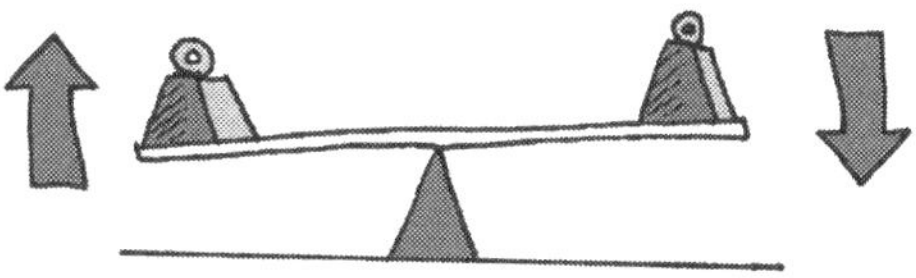

Back muscles contract with a balancing force

When you say...

Intertransversarii create tilt

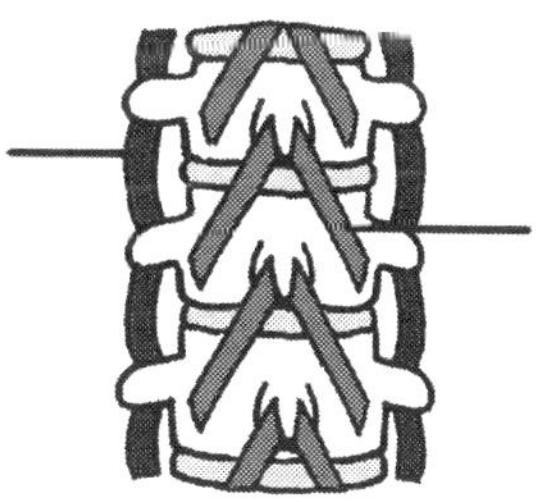

Multifidus and Rotatores create rotation

These muscles are not under your voluntary control. The cerebellum gives them orders in timely coordination with your actions.

Differences in Tilt Dynamics

Tilting to the right

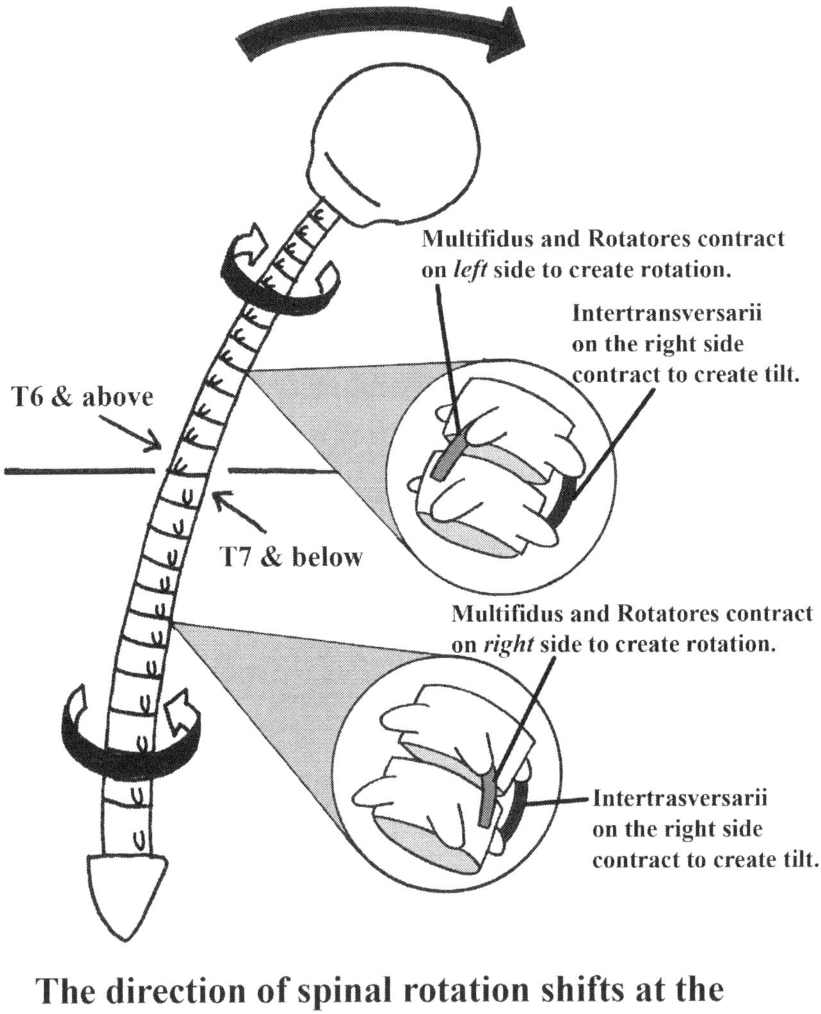

The direction of spinal rotation shifts at the level of T6-T7

Distribution of Movement Receptors

The greater number of receptors in our neck and upper back contribute proportionately more input to the cerebellum. This input has an impact on controlling everything below.

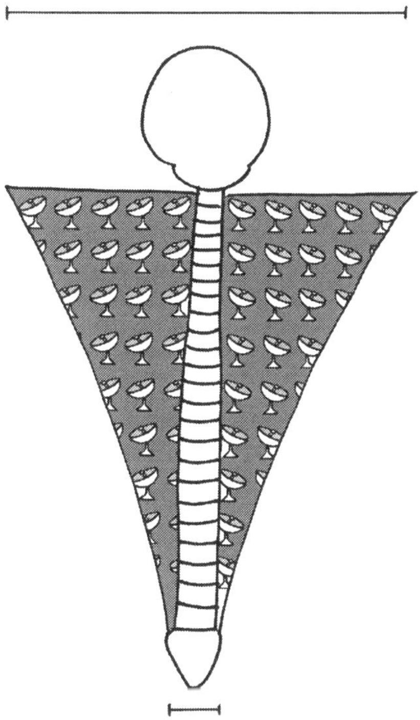

Therefore, the care of the upper spine is extremely important. While not obvious, low back pain can have its root cause "upstairs".

As above, so below...

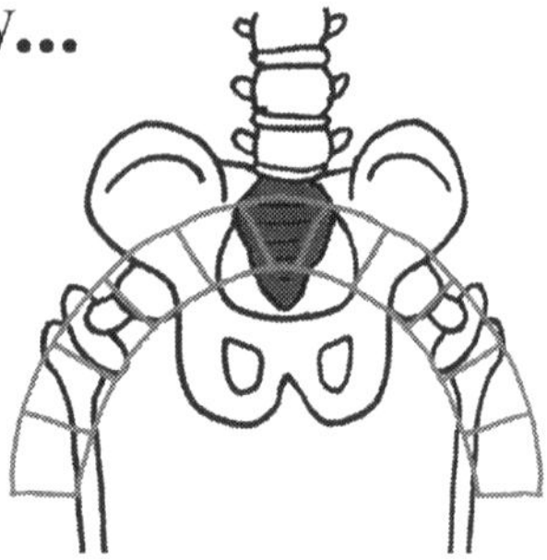

Just as a keystone is critical in holding an arch together, so too your sacrum is critical in maintaining structural stability from your hips down.

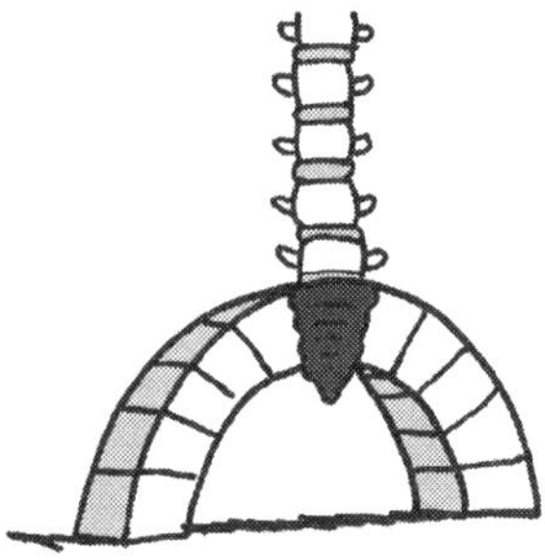

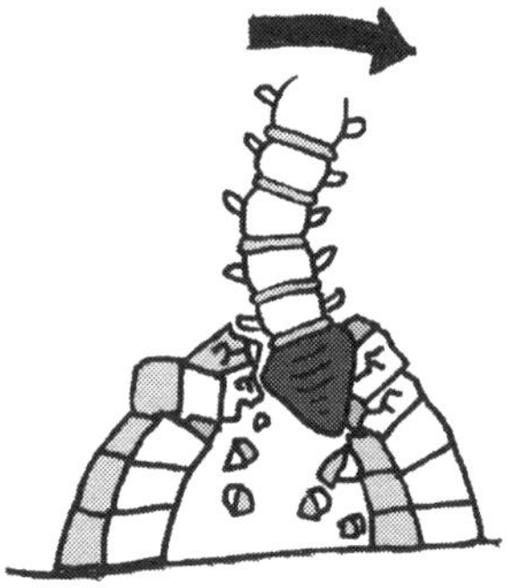

Unbalanced forces above will inevitably have a negative effect down below. In this way the sacrum is a bit like a drain trap in a sink.

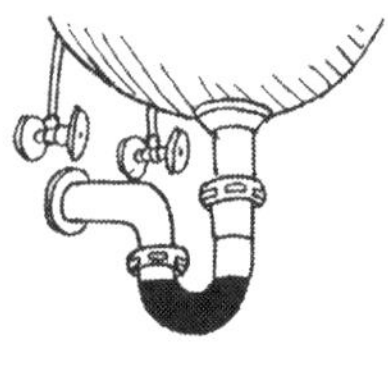

Problems above will often find their way down to the bottom causing lower back and leg pain.

TAKING CARE OF YOUR DISCS

Your spine is designed with a series of naturally occuring curves that serve to distribute weight downward.

With proper alignment weight is borne along interlocking facets towards the rear of each vertebra.

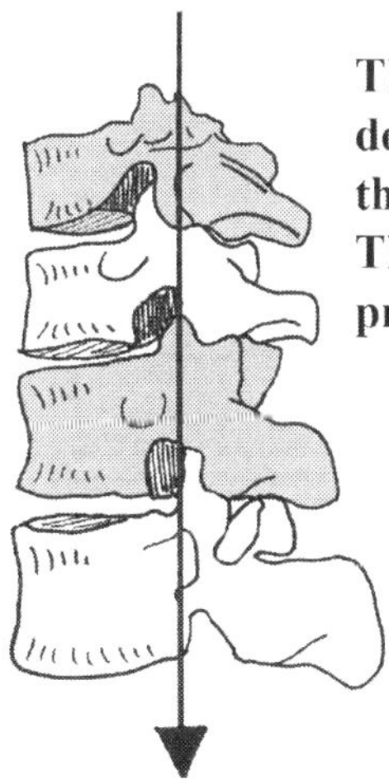

This is the ideal condition. What you don't want to do is put pressure on the front of the vertebra. This is when disc problems arise.

We want weight on the back of joints for activation of movement receptors.

Your discs act as shock absorbers between each vertebra in your spine. When disc problems occur it is a symptom of deeper problems.

Remember that immobilized muscle starts losing protein within 6-10 days.

Weakened back muscles contribute to loss of natural curvature, allowing pressure on the front of the vertebra and thereby compressing the shock absorbing disc.

Restoring range of motion to the joint to increase mechanoreceptor activation will reengage communication with the brain. The brain will activate the muscles. The strengthening of the muscle is the first step in restoring natural curvatures and avoiding more disc problems.

"The Eyes Have It!"

Yes. The muscles that control eye movement are wired into the same midbrain area that controls back muscles. Eye exercises can actually improve spinal range of motion.

Midbrain

Eye Muscles = Interconncted - Back Muscles

The cerebellum is critical in assuring that proper timing is part of the coordination.

MEANWHILE, BACK ON THE MOON...

Things are looking up after a series of adjustments to restore range of motion in the locked up areas of the spine...

Combined with diligent exercise to strengthen newly awakened back muscles...

Allowing for all the activities you love...

And the occasional rock, too.

Now it's time to take your health into your own hands. The following warm up exercises are designed to precede more advanced exercises, but if they are all you ever do you will most certainly benefit from it.

Future exercise routines will require a large exercise ball, a smaller ball, elastic resistance bands and special FTP balls.

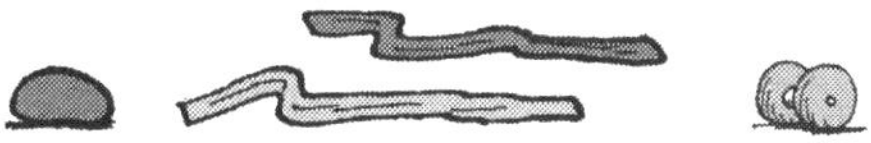

For now all you need is a small semi-inflated ball and a set of FTP balls for use in the Fulcrum Torque Procedure, a unique exercise that works the deepest muscles all the way up your back in order to improve range of motion throughout.

FTP balls

WARM UPS

1-ANKLE/KNEE ROTATIONS
With bent legs and hands on knees swivel around your ankles 10X in both directions.

2-HIP & LOW BACK ROTATION
With legs apart and hands on hips swivel an invisible Hula Hoop around 10X in both directions.

3-HEAD & NECK ROTATION
Standing erect, slowly swivel your neck around 10X in both directions.

4-COMBINATION PURSUIT & VOR*

With your head and eyes moving together trace the triangle shape 10X in both directions.

5-PURE VOR

With hands together overlap thumbs and, staying focused on thumb, make large circular motions with head 10X each way.

6-VISUAL PURSUIT WITH BODY MOTION

Knees bent, trace figure 8 side to side, following with eyes and head 10X in both directions. (Keep your knees behind your toes)

7-VISUAL PURSUIT WITH CONVERGING & DIVERGING

Trace figure 8 in & out while tracking your thumb with your eyes only 10X in both directions.

*Vestibular Ocular Reflex

8 - CROSS CRAWLS

A Cross Crawl cycle entails a left/right move in each plane of direction performed 10X. Until you become familiar with each movement you might want to concentrate on one move at a time.

As if marching in place, raise your opposite arm and leg

Lift your leg, crossing in front at a 45 degree angle as you cross opposite arm

Lift your leg outward at a 45 degree angle as you raise opposite arm

Lift your opposite arm and leg out to the side

Lift your opposite arm and leg, extending your leg straight back

If balance is a concern approximate moves keeping your feet on the floor.

Using the right leg as a reference point from an overhead view note the five different planes involved in this particular exercise...

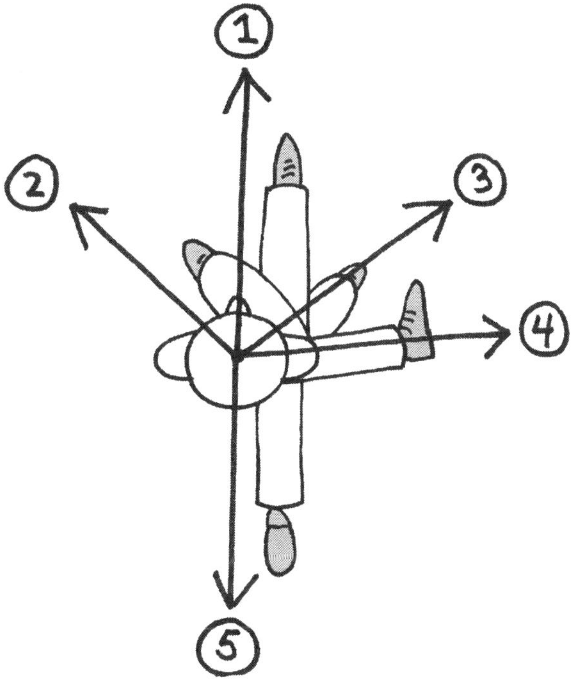

9 - ABDOMINAL HOLLOWING AND CROSS EXTENSION

Get down on all fours. Then, while sucking your tummy in, extend your opposite arm and leg and hold for 10 seconds. 5X on each side.

10 - MODIFIED NECK BRIDGES

Lie down with your head on a semi-inflated small ball. Press down, arching your neck 5X. Roll onto your left side and repeat, pressing down 5X. Once again on your back and finally by rolling onto your right side.

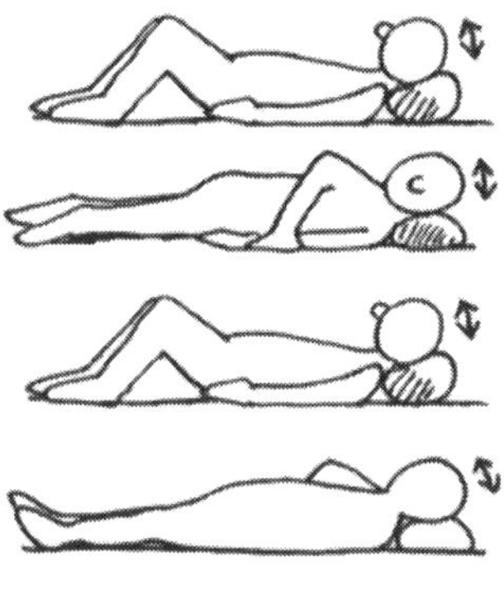

11 - FULCRUM TORQUE PROCEDURE

Requires FTP Balls

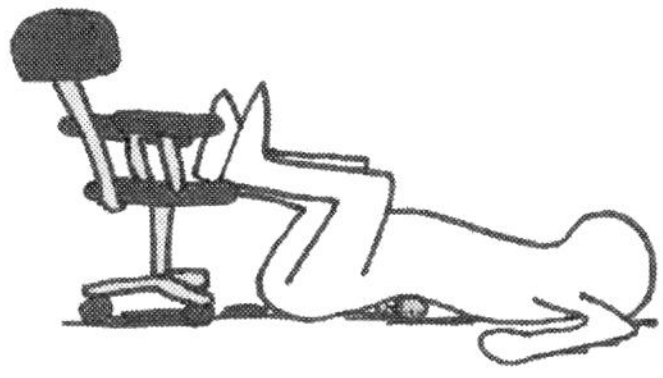

Lie down on your back with your feet elevated. Position the FTP Balls at the base of your spine. Do 5 reps of motion at each vertabrae as you work your way up your back.

Start by alternately reaching with your feet in a left-right combo so as to create a swivel in your hips 5X. Then move the balls up one vertebrae and repeat until you reach the area between your shoulder blades

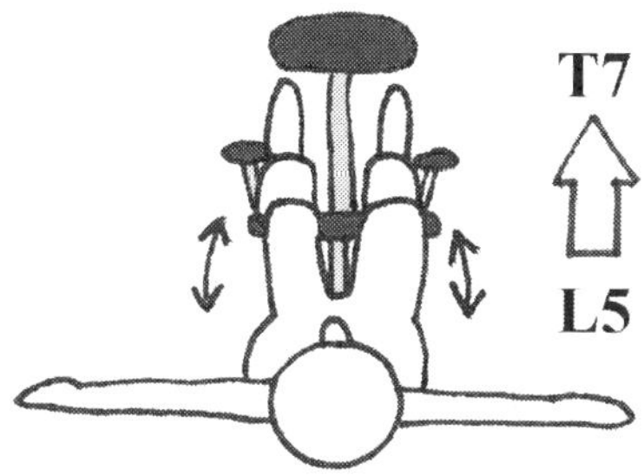

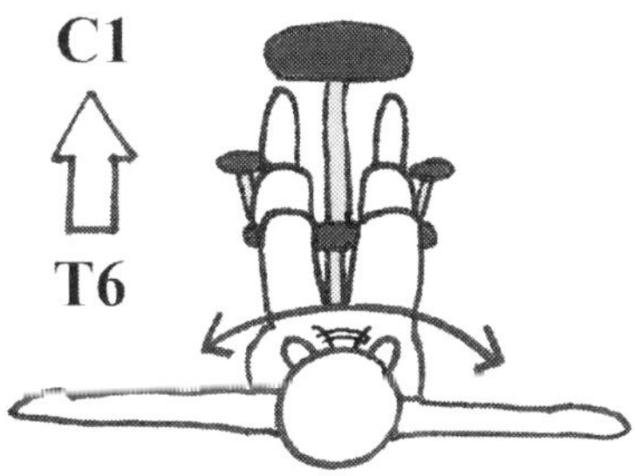

At this point switch your motion to a left-right head swivel, continuing up your back, swiveling 5X at each vertebae.

The differing motions are related to the different tilt dynamics that occur above and below T6-T7.

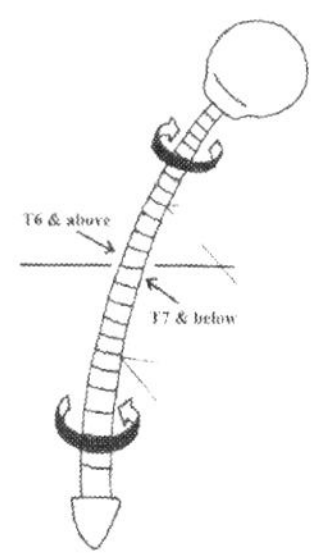

12 - SUPINE CROSS CRAWL CO-CONTRACTIONS

The final warm up mimics the first cross crawl position, but done while lying down.

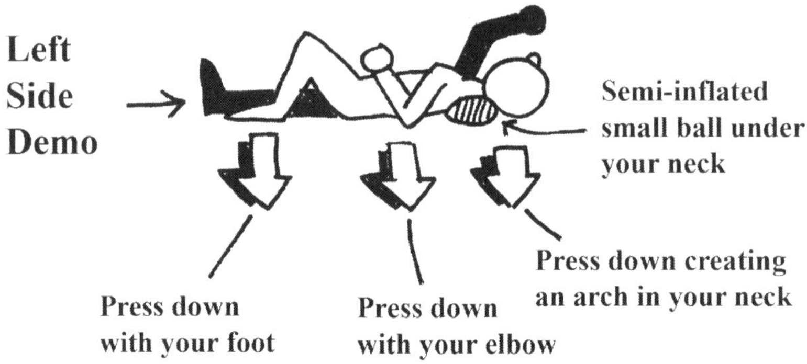

Press foot, elbow and head down simultaneously
Hold for 5 seconds, 5X on each side

This exercise will help to strenghen your deep back muscles from top to bottom.

It's one exercise you can even do in bed.

While you're still breathing there's always room for improvement...

Your health is a team effort...

You'll find the captain in the mirror.

APPENDIX
THE ANALOGY OF THE SPINE AS A GATE

The spinal column in humans is composed of seven cervical vertebraes on top of twelve thoracic vertebrae that rest on five lumbar vertebrae, which rest on the sacrum that is composed of five fused segments.

Each spinal joint articulates with the vertebra above, primarily through what's called a three-joint complex consisting of two facet joints and intervertebral disc. Vertebrae are held together by ligaments and the intervertebral discs. The individual vertebrae are moved in a coupled fashion of tilt and rotation by the coordinated efforts of spinal muscles. The spinal joints have ideal motions that each individual joint should perform. Specific muscles are engaged in the process of moving a vertebra in a particular direction.

The spinal joint complex may be compared to a common gate. When looking at a common gate, there is the gate post to which the gate is attached with hinges. When comparing the gate to the spine, the post and side of the gate would be consistent with the adjacent vertebra. The hinge on that gate would be analogous to the attachment of joint surfaces. The hinge component with the associated hinge screws are similar to the ligaments which would be binding the spinal joint complex. An individual providing the physical power to open and close the gate through its range would be analogous to the muscles working around the joint.

Head over to www.theneurotechnologies.com to read more about The Spine As a Gate...

Also check out Book One!!

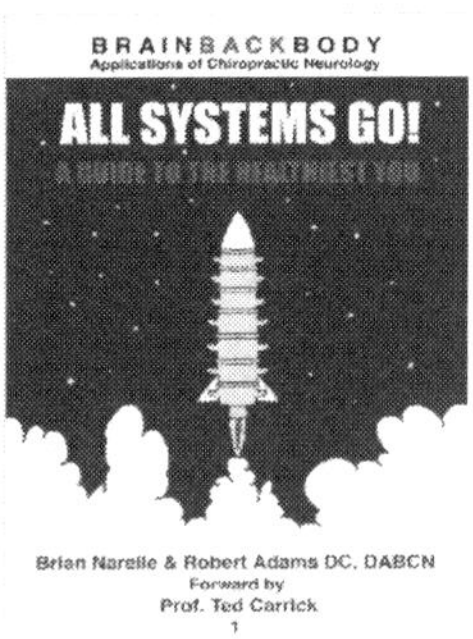

Brian Narelle began cartooning at the age of eight. No one ever told him to stop, so he didn't. He specializes in making complicated things digestible, be it IBM technology or emergency medicine, and has illustrated many books. As a writer he started out on Sesame Street and has gone on to write award winning PBS Specials and dozens of educational film. Coming full circle, he currently teaches cartooning to eight-year-olds at the Charles Schulz Museum in California.

Brian Narelle
briannarelle@comcast.net
Narellecreative.com
707.586.2070

Made in the USA
Monee, IL
11 September 2026

62770614R00020